FLOWERS
of Hawaii

Written by Allan Seiden

AN ISLAND ❀ HERITAGE BOOK

*F*LOWERS of HAWAII

An abundance of exotic flowers provide Hawaii with a tropically colorful beauty year round. It wasn't always that way, for the thousands of miles of open seas that surround the islands isolated them from the rapid introduction of flowering plants. Before man settled Hawaii, only the hardiest of seeds, spores, nuts and fruits survived water-borne, bird-carried or wind-blown journeys that brought them from distant homelands to Hawaii's volcanic shores.

The vast majority of the thousands of flowering plants that now give Hawaii the appearance of a botanical garden have actually been introduced within the past century. Their origins range the globe, with Polynesia, North and South America, Asia, Africa and the world's tropical islands each providing a share. The result is a landscape alive with an unusually diverse selection of flowering trees, vines and shrubs.

Over the years Hawaii's floral inheritance has indeed been enhanced, making the islands all the more beautiful. Let this book be your guide in discovering the flowering abundance of contemporary Hawaii.

◄ Hybrid Hibiscus

Hibiscus

Hawaii's State Flower

Hundreds of diverse and magnificent hybrid varieties of this hardy tropical shrub have become established in Hawaii, where several native species prospered even before the arrival of the Polynesians. Blooming throughout the year, hibiscus flowers will remain fresh and unwilted for a full day after they are picked. Since they crush easily and their color may stain clothing, they are not used for leis. Called *koki'o* in Hawaiian, this easily grown shrub produces flowers in a multitude of colors and color combinations. The yellows were probably hybrids when brought to Hawaii. Crossed with other hibiscus, its offspring range from pale yellow to bright yellow, from orange to orange-red. Most yellows are large, and some feature double blossoms.

Orchids

One of the most abundant and varied of all plant species, the orchid is also one of the most beautiful. Today tens of thousands of hybrids have been created from a worldwide selection of indigenous forms, including a number of small orchids native to the Hawaiian Islands.

VANDA ORCHID
MISS JOAQUIN HYBRID

While orchids are grown on all islands, the Big Island and Oahu are the primary commercial growers. The pink and purple orchids that are found in leis, maitais and a hundred other places are almost all the Miss Joaquin hybrid, which flowers abundantly throughout the year.

CATTLEYA

Remember the orchid corsage? It was no doubt a Cattleya, the most showy of all the orchid types.

CATTLEYA HYBRID

An elegant cattleya hybrid reveals the reasons that orchids have long captured the imagination and been idealized for their beauty.

PHAELENOPSIS

A wide-eared Phaelenopsis white is yet another variation on the orchid theme.

EPIDENDRUM

Like hovering ballerinas, a collection of tiny Epidendrum orchids are a less well-known member of the orchid family.

◄ DENDROBIUM

A cluster of white flowers bursts forth from a single dendrobium stem.

ANTHURIUM

Among the most popular of Hawaii's tropical flowers is the exotic heart-shaped Anthurium. It ranges in color from white to pink, to a deep, rich red, and occasionally green or green and white. The finger-like spike is white, pink, or yellow. Its true flowers are found on the spike. The heart-shaped bract is a modified leaf; it has a waxy quality and is often mistaken for artificial. Once cut, the flowers can last up to three weeks in water.

BIRD OF PARADISE

Colorful "plumage" and a beak-shaped sheath make the Bird of Paradise one of the more exotic and popular of Hawaii's floral transplants. This South African member of the banana family features brilliant orange petals and dark blue stamen that rise from a cluster of tall stems that blossom consecutively over a ten to twelve day period.

Lilies

SPIDER LILY

This native of tropical America blooms year round in a spidery cluster of white and yellow that justifies its name. Six thin petals and six stamen extending from the center of the flower define it as a member of the lily family.

WATER LILY

India and Africa provided Hawaii with Water Lilies. Today they are grown in ornamental pools in colors ranging from white to yellow, lavender blue to pink purple.

PASSION FLOWER

The Passion Flower's Hawaiian name, *lilikoi,* derives from the place on Maui where it was first planted. Today a wide variety of these elegant flowering vines can be found in Hawaii, primarily in shaded forests where they hang from trees or sprawl over walls and rocky terrain. The ripe yellow fruit they produce has a tart taste and is used in jams, jellies, and juices.

CHENILLE PLANT

The flowers of this tropical plant form thin furry tails that grow to eighteen inches in length. This native of India actually has no petals, with the flower formed by the budding of its elaborate stamen.

CUP OF GOLD

A Mexican branch of the tomato family, the giant blossoms of the Cup of Gold average nine inches in diameter. Their individual impact is enhanced by the abundance of blossoms that appear at one time. Flowering occurs from January through March, with the heavy waxen buds opening so quickly that it can be seen as the flower unfolds. A rarer white-blooming variety is also sometimes seen.

YELLOW ALLAMANDA ▲

Called *lani-alii,* or Heavenly Chief, by the Hawaiians, this plant grows either as a vine or a shrub. As a shrub it is commonly seen as a hedge. Two or three flowers blossom at a time from a cluster of buds that are surrounded by waxy green leaves. In its native Brazil the Yellow Allamanda lacks the brown streaks that now identify its Hawaiian offshoot.

FUCHSIA

Originally brought in from the U.S. mainland, the Fuchsia is technically a weed. It is now found in cooler, higher mountainous areas around the islands, both in gardens and in the wild.

BOUGAINVILLEA

Brazil has provided Hawaii with the Bougainvillea, one of its most common flowering plants. This densely flowering vine also grows as a shrub. It flowers throughout the year, bursting to life with masses of purple, crimson, orange, white and pink flowers. The color is actually provided by modified leaves, called bracts, that cluster around the plant's tiny, white, almost inconspicuous flowers.

Gingers

KAHILI GINGER

This native of the Himalayas takes its Hawaiian name, *kahili,* from the tall standard carried before Hawaii's chiefs as a sign of royalty when they appeared in public. The blossoming Kahili bears a resemblance to the feathery kahilis of old, with their lineup of yellow flowers and red filaments emerging from a fibrous stalk.

TORCH GINGER

The waxen cone of the Torch Ginger grows at the end of a long stalk that grows directly from the ground amidst encompassing leaves. Small white flowers emerge from between the scale-like red or pink bracts that provide this ginger species with its unusual appearance.

WHITE GINGER

Delicacy and a heady fragrance describe the White Ginger, a flower found in abundance in upcountry fields where they flower in acres-large colonies.

◀ ROYAL POINCIANA/ FLAMBOYANT

A graceful transplant originally from Madagascar, the Royal Poinciana can grow up to forty feet tall. Flowering occurs May through August, when the tree temporarily loses its small fern-like leaves. The result is an inspiring dome of red and orange flowers that makes this tree a popular choice for public parks.

AFRICAN TULIP

This graceful tree, a native of tropical Africa, blossoms throughout the year. Its rich orange-red flowers appear on thickly-leaved branches, adding their exotic brilliance to the rich greens of the lowland forest. Individual blossoms open at the edge of a cluster of buds that is constantly maturing. The flower is followed by a seed pod that may be up to two feet long.

Plumeria

This most common of lei flowers is called frangipani outside of Hawaii. A native of tropical America, durability, sweet fragrance, and an abundance of flowers that bloom throughout the year are behind its popularity. Velvety flowers range from white and yellow combinations to deep pink with a cerise center. Various hybrids add to its natural range. The Plumeria's sticky white sap can be poisonous.

Heliconia

LOBSTER CLAW

These relatives of the banana produce a flowering stem of bracts from which small green flowers emerge. The various heliconia species all feature large leaves, some of which may grow up to twenty feet in length. The Lobster Claw is most commonly found in the Hawaiian rain forest in the shade of overhanging vegetation.

PARROT'S BEAK

The exotic, almost artificial looking Parrot's Beak Heliconia is similar in shape to the Lobster Claw. This early spring bloom makes spectacular flower arrangements that last weeks.

DUCHESS PROTEA

The Duchess grows on a small bush, producing a blossom whose central crown is surrounded by a ring of red-tipped petals.

NIGHTBLOOMING CEREUS

The giant breathtakingly beautiful white and yellow flowers of this Mexican cactus bloom only at night between June and October. Brought to Hawaii from Acapulco by an amply impressed sea captain, these tenacious climbers cover walls, fences and tree trunks in clusters of exquisite blossoms. Buds develop in cycles, with flowering of the buds of a particular cycle occurring all in one night. Ten to twelve days later a new flowering cycle will begin. The flowers wilt and close with the morning heat.

SILVERSWORD

Found only on the upper slopes of Maui's Haleakala and on the high peaks of the Big Island, this member of the sunflower family grows among rocks or in dry cindery soil. In immature plants, silvery-green leaves covered in wisps of soft "hairs", are its main attraction. A collection of flowers grows on a two to six foot stem that rises from this leafy core. As the flowers bloom, the leaves die. Once seeds are produced, the parent plant itself dies.

◀ STEPHANOTIS

In Hawaiian this flower is called *pua male,* or "marriage flower". These pure white flowers are waxy and have a delicate fragrance. They are used in wedding bouquets and leis.

ORCHID TREE

Despite its name and the orchid-like appearance of its flowers, the Orchid Tree is no relative of its floral namesake. Deep green, moth-shaped leaves surround both the lavendar and white floral variations of this species which originated in India.

CANNA

Found growing wild and in Hawaii's gardens, the Canna originated in tropical America's rain forests. Introduced soon after Hawaii's discovery by European explorers, it grows from three to five feet tall, sporting green or purple leaves and velvety flowers that range from pink to scarlet red, with additional yellow varieties.

OHIA LEHUA

A native of Hawaii's upcountry forests, the *Ohia Lehua* is one of the first trees to establish itself on newly formed volcanic terrain, rooting in soil that accumulates in cracks in lava-formed rock. An extremely varied species, the Lehua flower is most often a red pompom, although pink, yellow and creamy white variations are also found. Pollenated by a variety of native insects and birds, the flower is hardier than its delicate appearance reveals.

POINSETTIA

Discovered in Mexico in 1828 and brought to Hawaii soon after, the Poinsettia has since become a tropical favorite. A November or December visit to Hawaii is likely to find them in floral abundance, either by the roadside or in ornamental plantings. Hawaii is home to red, pink and cream white variations.

IXORA

A cluster of individual flowers form a pompom of color on the flowering Ixora. This relative of the coffee tree is originally from the East Indies. Its flowers are occasionally used for specialty leis that are wound in spirals. A fragrant white is also found in Hawaii.

CORAL SHOWER

The first of the shower trees to bloom each year (March through May), it is also the largest. This native of tropical America can be covered in a dense profusion of variegated petals. Flowering is in clusters that grow from stems on its branches.

OLEANDER

These tall shrubs from Asia Minor and India are members of the periwinkle family. Commonly used as hedges or in decorative plantings, their flowers range from white to cream, from pink to red. Their beauty not withstanding, the oleander is a poisonous plant. Even its fragrance can cause sickness in an enclosed space. In Hawaii children are warned against using Oleander sticks for toasting marshmallows or as barbecue skewers.

CHRISTMAS-BERRY

This plant is often seen in Hawaii
growing wild from seeds dropped
by birds. It bears bright red berries
in thick clusters from fall through
mid-winter. The berries decorate
many a wreath and flower arrange-
ment much as holly is used in
other parts of the world during the
holiday season.